45 4068493 8

D1151429

Natacha's Animal Rescue

Rob Waring, *Series Editor*

HEINLE
CENGAGE Learning

Australia • Brazil • Japan • Korea • Mexico • Singapore • Spain • United Kingdom • United States

Words to Know

This story is set in southern Africa, in the country of Namibia. It takes place in a remote region called Fish River Canyon.

 Animal Rescue Mission. Read the paragraph. Then match each word or phrase with the correct definition.

Conservationist Natacha Bateau [nətɑːʃə bætoʊ] makes her home in a beautiful region called Fish River Canyon. In the past, so many wild animals lived there that some said the pristine wilderness area was almost like paradise. Then, hunters began killing many of the game animals for food or sport. Some species were totally wiped out so that none are left in the canyon. Natacha is now trying to bring these animals back to the canyon and Ulf Tubbesing [ʊlf tuːbəsɪŋ], a vet, is working with her to help keep the animals healthy.

1. canyon _____

2. pristine _____

3. paradise _____

4. game animal_____

5. wipe out _____

6. vet _____

a. clean and new; perfect

b. destroy until no more exist

c. a long, deep opening in the earth's surface

d. animals hunted for food or sport

e. a doctor for animals

f. a perfect place, often considered to be imaginary, where everything is beautiful and peaceful

B Animals in Africa.

Animals in Africa. Here are some animals you will find in the story. Label the picture with words from the box.

a black rhinoceros (rhino)	a giraffe	an oryx
cheetahs	a leopard	springboks

1. _____

2. _____

3. _____

4. _____

5. _____

6. _____

The vast region known as Fish River Canyon in southern Namibia is a huge open space with red, desert-like earth. The area was once home to a rich number of various big game animals such as rhinoceros, cheetah and oryx. In the past, the wildlife here was able to run freely through the strikingly beautiful landscape, but all of that has changed with the coming of modern times – and it hasn't changed for the better.

The last hundred years in particular have brought constantly spreading settlements and an endless supply of hunters to the area. Over the years, hunters, **poachers**[1] and farmers have hunted and killed virtually all of the canyon's animals for sport, for food or to make room for **livestock**.[2] Nowadays, very few of the original wild game animals remain in and around Fish River Canyon. They're more likely to be found on restaurant menus or on walls as sporting **trophies**[3] than racing across the canyon floor.

Fortunately for Fish River Canyon, there's hope. Conservationist Natacha Bateau has a dream. Her greatest desire is to bring these animals back to their natural habitat and to allow them to freely move about the pristine, wild territory of the canyon once again.

---

[1]**poacher:** a hunter who takes game animals illegally
[2]**livestock:** farm animals, such as cows and sheep
[3]**trophy:** an animal that has been hunted and stuffed for display

WEST HERTS COLLEGE
CONTROL 81456
ITEM 4 54068493 8
CLASS 428.6 NAT 6.2
DATE 18/11/11

Skim for Gist

Read through the entire book quickly to answer the questions.

1. What is the reader basically about?

2. Give a general review of what happens in the story.

Natacha was born and raised in Paris, but moved to the wilds of Fish River Canyon in 1995.

A native of Paris, France, Natacha came to live alone in Fish River Canyon several years ago and the canyon immediately found a place in her heart. She thought it was a delightful place to live, almost a paradise and she fell deeply in love with it. Though she had grown up in an urban environment, Natacha had always wanted to live in the wilderness. As a child she would often dream of living among the animals somewhere in a wild land far, far away. Coming to Africa was a realisation of that childhood dream.

Fish River Canyon was far from perfect, though. By the time Natacha arrived, there were almost no wild animals left in the area. 'When I first came here, the game animals had been wiped out by man,' she explains 'and most of their **predators**[4] also. So there were virtually no animals left.' It may have been a paradise at one point in its history, but it was certainly in need of some help to bring it back to its ideal state. Natacha Bateau was ready to meet the challenge.

---

[4]**predator:** an animal that kills and eats other animals

Many people talk about restoring the wilderness, but Natacha doesn't just talk; she actively does something about it and the case is no different with Fish River Canyon. Her plan is to bring nature back into balance by rescuing game animals in other parts of Africa and bringing them to Fish River Canyon.

Natacha, usually accompanied on her drives across the countryside by her playful leopard companion, Chemun, admires and respects animals. She's presently engaged in an ongoing commitment to repopulate the lands near her home with the animals that used to live in the region. She feels that it's critical to relocate animals here so that the canyon can be restored to the paradise that it once was. 'To me it's important to relocate animals here because I would like to see this land go back to its natural state,' she explains. However, such a project isn't going to be exactly straightforward. There will be a number of challenges for the strong-willed conservationist to overcome.

Natacha is well aware that careful planning is going to be required to bring back the animals to a place like the canyon. It will also be extremely hard work and a task that she simply cannot do alone. For this complicated project, she's going to need **allies**,[5] people who are equally as passionate about animals as she is. Fortunately, she's surrounded by others who want to help her to achieve her dream. Ulf Tubbesing is one such person.

As one of Namibia's best vets, Ulf shares Natacha's **devotion**[6] to saving game animals. When Natacha met him, he was protecting wild animals from being killed by hunters and farmers and caring for animals that needed new homes. Like Natacha, he's a great animal supporter, spending his days in the constant company of animals. He even has a playful cheetah cub sitting with him sometimes while he works at his computer. It didn't take long for the two animal advocates to discover that they have compatible goals, and soon they began to combine their efforts in order to help save animals and repopulate the region.

---

[5]**ally:** a partner or friend, especially in a difficult situation
[6]**devotion:** dedication; loyalty

When Ulf discusses Natacha and her home, it's obvious that he understands exactly what she's trying to accomplish. 'Well, Natacha bought the farm some five years ago,' he explains 'and she's been dreaming about building this thing up into – or **reverting**[7] it back to – a natural state, the way it used to be a hundred years ago before man interfered.'

Ulf describes what he and Natacha plan to do in this pristine area with its delicate ecology. 'This is a very, very pristine and special area. It's a very sensitive habitat, a very sensitive ecology, [with] very, very balanced plant life. And our entire aim,' he says, 'is to just re-establish that ecology and that **ecosystem**[8] as it used to be.' Repopulating a place like the canyon will require excellent preparation and intense planning, though and Natacha and Ulf must implement their plans carefully.

---

[7]**revert:** return to an original or previous state
[8]**ecosystem:** the balance between living things
and their environment

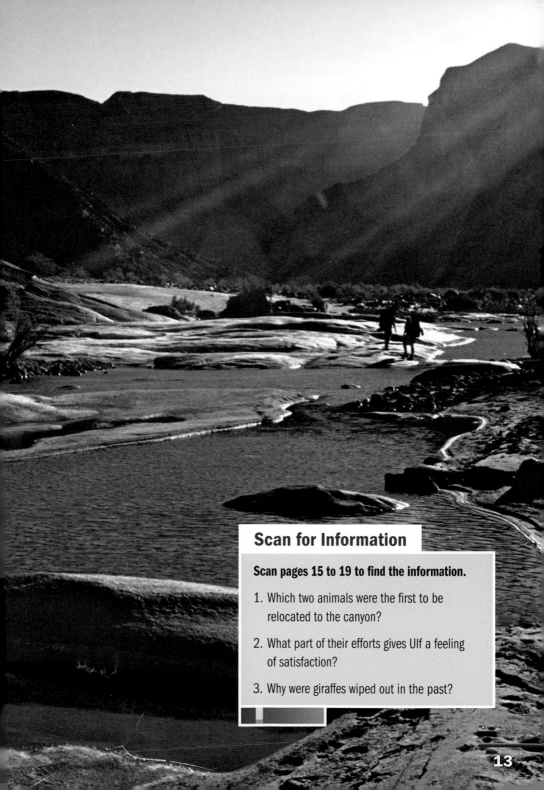

Scan for Information

Scan pages 15 to 19 to find the information.

1. Which two animals were the first to be relocated to the canyon?

2. What part of their efforts gives Ulf a feeling of satisfaction?

3. Why were giraffes wiped out in the past?

The repopulation of Fish River Canyon is actually being done in carefully planned stages. The first animals relocated by Ulf and Natacha were springboks. As the animals were released into the dry, desert countryside, they sprang from their cages, delighted to be free again and soon escaped into the wilderness. Later, the two conservationists captured and relocated a cheetah family to the area near Natacha's ranch. The animals were brought to the area in cages and were obviously uncertain about the move. They hesitatingly left the safety of their temporary homes behind them to face the unknown. After a short period of adjustment, however, the cheetahs likewise ran off into the canyon. As the sun set in the distance, it became apparent that they would soon adapt to their new environment.

a springbok

For Ulf, there's always a tremendous feeling of satisfaction when animals are released from their cages and set free into this magnificent landscape. To be personally involved with it all brings him great joy. He describes his feelings, 'It's a very exciting project ... to come back to a place year after year and start seeing more and more animals **roaming around**[9] and having the sensation that just about every wild animal that you see, we put in here. You ... touched them all with your own hands. You physically worked hard on [catching] them to get them here. It's great.'

But, no matter how exciting they were, those early relocations were rather simple when compared with what Ulf and Natacha plan to do next. Their newest challenge is to try to capture and transport the world's tallest animals: giraffes.

[9]**roam around:** run around freely

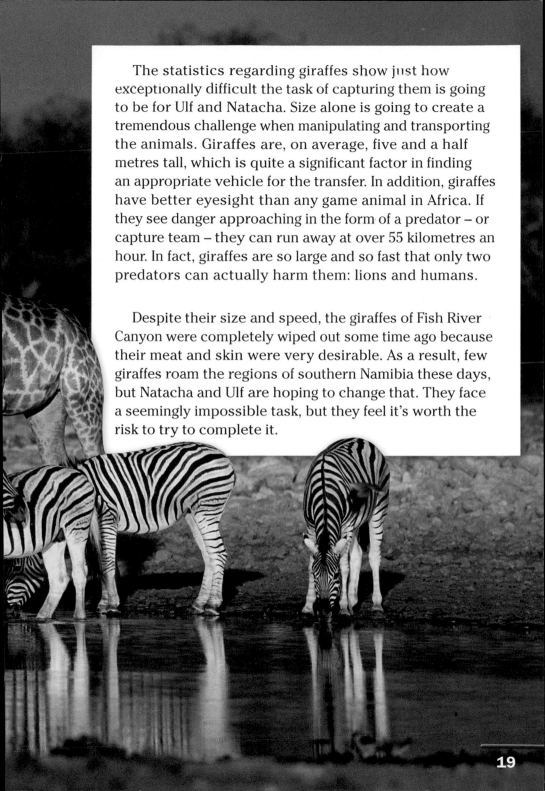

The statistics regarding giraffes show just how exceptionally difficult the task of capturing them is going to be for Ulf and Natacha. Size alone is going to create a tremendous challenge when manipulating and transporting the animals. Giraffes are, on average, five and a half metres tall, which is quite a significant factor in finding an appropriate vehicle for the transfer. In addition, giraffes have better eyesight than any game animal in Africa. If they see danger approaching in the form of a predator – or capture team – they can run away at over 55 kilometres an hour. In fact, giraffes are so large and so fast that only two predators can actually harm them: lions and humans.

Despite their size and speed, the giraffes of Fish River Canyon were completely wiped out some time ago because their meat and skin were very desirable. As a result, few giraffes roam the regions of southern Namibia these days, but Natacha and Ulf are hoping to change that. They face a seemingly impossible task, but they feel it's worth the risk to try to complete it.

While Fish River Canyon is suffering from a shortage of giraffes, more than 650 kilometres away in the northern region of Namibia, the number of giraffes is growing quickly. In fact, in some areas the giraffes have run out of room and must live in overcrowded game parks where people pay to see them. Because there are simply too many of them, they've become problem animals in some cases. Apart from killing them, the only solution is to move the giraffes to remote areas. This is where Natacha and Ulf may be able to help.

By taking the giraffes from the north and relocating them to Fish River Canyon, Natacha and Ulf may actually be able to help the animals. It may be the ideal solution to the overcrowding problem – and just what Natacha and Ulf want. However, it still remains to be seen if they can manage the one great challenge to their plan: capturing these great animals and transporting them back to Natacha's land safely.

There's a great deal of preliminary preparation that has to be made before the giraffes can be brought south. Natacha drives into the countryside in her truck, accompanied as always by Chemun. Here, she carefully examines the area for potential places where they can release the giraffes in Fish River Canyon. This thorough search for the perfect spot happens before each rescue and it's crucial to the success of the project. She explains what criteria must be met in finding and choosing the right place. 'Now we're looking for the right spot to release the giraffes,' she says, 'and the elements to look for are accessibility … and the availability of water – not very far [away] – [and] trees.'

Finally, after driving around the area for some time, she finds what she thinks might be the right spot for the giraffes. She's found a place with enough space and trees, which make it an area where the newly moved animals could exist comfortably. She stops her vehicle and inspects the spot up close; it appears to be just what she's looking for. The first task completed, Natacha must now begin thinking about completing the team that will be helping her with the forthcoming rescue.

For help with her plan to capture and release the giraffes, Natasha goes to see another ally. This time it's vet Hermann Scherer. In Namibia, Hermann is the man to see when one wants to catch a giraffe and he's able to show Natacha and Ulf exactly what he can do to help. He's a superb **marksman**[10] and his tremendous skill with a gun has allowed him to specialise in capturing live animals in the wild. More importantly, Hermann's speciality is catching giraffes, not the easiest of tasks and one that requires great expertise. This combination of skills is going to make Hermann a valuable member of the group, and he'll be bringing a group of assistants with him as well. Now, with Hermann and his team, as well as Ulf, Natacha is ready to attempt the giraffe rescue and relocation plan for which she has been preparing so long.

---

[10] **marksman:** a person skilled at shooting guns

Predict

Answer the questions using information you know from reading to this point. Then, check your answers on page 27.

1. How will the team capture the giraffes?

2. How long will the team have to capture the animals?

3. In what type of vehicle will they move the giraffes to Fish River Canyon?

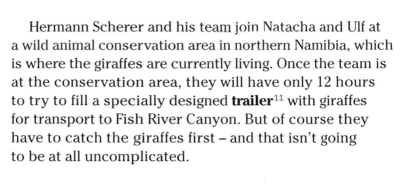

Hermann Scherer and his team join Natacha and Ulf at a wild animal conservation area in northern Namibia, which is where the giraffes are currently living. Once the team is at the conservation area, they will have only 12 hours to try to fill a specially designed **trailer**[11] with giraffes for transport to Fish River Canyon. But of course they have to catch the giraffes first – and that isn't going to be at all uncomplicated.

The plan commences with the team cruising around the conservation area in an attempt to find an appropriate animal. They need to carefully select which animals will be taken back to the canyon because the animals must not only be the right age and strong enough to be moved to a new home, but also able to survive being hunted and injected with powerful drugs to make them sleep. If there are complications during the capture, the giraffes can't stay asleep long as they could be at risk and could easily die. This means that the team will have only a few minutes to catch each giraffe and get it ready for transport before giving it another drug to make it wake up. The process will be far from easy, but once it's done, it will allow the animal to be safely transported by trailer to its new home.

---

[11] **trailer:** a large container that can be pulled by another vehicle

As expected, the main difficulty with the plan lies with the giraffe's exceptional eyesight and of course, its height and speed. Those advantages allow the animals to easily spot Hermann and the team from far away and enable them to keep at a safe distance. It's going to be difficult for Hermann to shoot the animals from a great distance, but luckily that won't stop this excellent marksman. Hermann, Ulf and Natacha continue their search for some time, seated on top of their trucks in order to have the best view of the landscape and increase the likelihood that one of them will see a giraffe.

After searching for some time, the team finally comes upon a pair of giraffes far out in the grasslands. The animals have not yet noticed the team, so Hermann takes advantage of the opportunity. He cautiously sets up his gun and trains the sights on a large male. He carefully squeezes the trigger and suddenly there's a quiet clicking sound as the gun fires a **dart**[12] full of a sleep-inducing drug into the giraffe. The animal immediately runs away, surprised and confused about what has happened. As the team carefully tracks his movements, Ulf announces that he is quite pleased with the results. 'That was a really nice hit,' he says quietly. 'Now we just have to wait and see. [We have to] wait for the dart to take effect.'

---

[12] **dart:** a small pointed object that sometimes contains a drug

The large male giraffe has now run a few hundred metres away so the team's trucks follow, thus beginning the chase to capture the animal. As the animal begins to slow, the truck pulls up beside him. Hermann's team is ready with specially tied ropes that are designed to go around the animal's neck to assist in slowing him down. The expert driver carefully maintains his speed while another team member slips the rope around the animal's head – they've got him!

Once the team has caught and restrained the animal, they must work quickly to ensure that the giraffe's enormous head remains upright. If its head falls down or bends sharply, the giraffe won't be able to breathe properly – and if he can't breathe, he could die. The operation requires effort from the whole team, as the animal's neck and head are quite heavy. Two men hold the animal's head up while Natacha strokes its nose to calm the animal. Meanwhile, Ulf rushes in to check that the animal is okay. It's a race against time to ensure that everything goes smoothly, and that the animal is not put at risk. There are anxious moments while everyone wonders if the giraffe will survive this stressful operation.

Once the team has ensured that the 900-kilogram animal has been positioned correctly, Ulf gives the animal the other drug in order to wake it up. Again, they have to work hurriedly as the giraffe will soon **regain his senses**.[13] Once that occurs, the huge animal will likely become alarmed once again and could be a danger to anyone near him. Ulf gently pours water over the giraffe's head to make sure that he doesn't become too hot. They then cover the animal's ears with wraps and put a mask over his head so that he can't hear or see anything, which will keep him calm as he becomes conscious.

Finally, the huge animal begins to wake up, slowly realising what is happening around him, but still confused. The men tie ropes to the animal's neck and legs in order to help guide him to the trailer. It takes several men to get the animal safely to its transport, but at last the seemingly impossible task is completed. As they finish their work and settle the tall animal safely into the trailer, Hermann takes a moment and turns to say, 'Natacha, congratulations! You are the owner of one **bull**.'[14] Natacha is absolutely thrilled with the results and thanks the expert shot for his fine work. The race against time has worked – at least for this giraffe.

_____

[13] **regain [one's] senses:** become conscious
[14] **bull:** the male of certain types of animals

ear wrap

mask

dart

OVERSIZ

DENEFFE SIGNS CLASS 2 AS 1906

The team proceeds with their plan and manages to catch a second giraffe using the same technique. At the end of a long and challenging day, the team has successfully caught not one, but two giraffes; however their job isn't over yet. They must now begin the 800-kilometre drive to Fish River Canyon, which involves driving all day and all night in order to get there as quickly as possible. They are forced to keep moving in order to limit the time that the giraffes must spend in the heat and confinement of the trailer. The less time spent in the trailer, the more likely it is that the animals will survive the trip.

Finally, after a nonstop journey of 20 hours, over extremely rough roads, the team and the giraffes arrive at the spot previously chosen by Natacha. Natacha guides Hermann and Ulf to the place where they can release the giraffes. 'This is the spot … near those big trees,' she says indicating the direction of the animal's new home. Ulf agrees with Natacha's choice, 'There's a lot of moisture in the leaves. You can see how nice and **lush**[15] [and] green the trees are.' The place has everything the animals need: life-sustaining water, trees for food and perhaps most importantly, no guns.

Next, the most delicate – and tense – part of the rescue happens. The team opens the door of the tall trailer to release the animals. The first giraffe hesitates before leaving the safety of an environment that it knows, but finally it appears at the door, cautiously looking around the unfamiliar territory. It seems reluctant to come out and the one thought on everyone's mind is: will the giraffe take that first big step to freedom?

---

[15] **lush:** green and full of life

After a few tense moments, the giraffe walks out of the trailer, uncertain at first, but then with more confidence. Finally, he's outside in the bright light of the desert landscape and he seems somewhat uncomfortable to be there. Little by little, the giraffe becomes more and more at ease in his surroundings. He looks around and after his first cautious steps, he is soon joined by the second giraffe. Within minutes the two animals run quickly away into their beautiful new surroundings, happy to be free once again, and not realising just how lucky they are to have their new home.

Natacha is understandably pleased with the results as she watches the giraffes. 'I'm always very happy that the animal comes here in good shape,' she reports. 'And when you can take them out of the truck one by one, you see them full of life and bouncing around and running … running away. That's wonderful. And you know that here they're going to have probably the best life they can have.'

Ulf is also excited about their achievement, and can't stop smiling as he watches the pair of giraffes run off into their new home. He talks about the experience and underlines the positive nature of the work he and Natacha are doing. 'It's the most incredible feeling,' he says. 'It's like you are so physically exhausted … you're totally **euphoric**.[16] It's a feeling of real achievement, of having done something positive.'

---

[16] **euphoric:** extremely happy

Besides removing the giraffes from an overpopulated area, Natacha also hopes for another positive outcome from the giraffe rescue: more giraffes. She looks forward to the giraffes that she has brought to Fish River Canyon breeding and adding to the giraffe population. Perhaps, she thinks, they'll be the beginning of a new giraffe herd in this area, like those that were lost so very long ago.

Although Natacha and Ulf are enthusiastic about the giraffe project's success thus far, they aren't planning on sitting back and doing nothing now that it's been completed. As the sun sets over the beautiful landscape that they call home, the two take time to discuss even more relocation projects that they have in mind.

Next, Natasha and Ulf are preparing to rescue yet another type of animal and bring it back to the canyon. This time they're planning to bring back the rare black rhinoceros. The black rhinoceros promises to be an especially important addition because, though there were once many of them inhabiting Fish River Canyon, there are now only a few black rhinos left in the world. In fact, the animal is so rare that it's in danger of becoming extinct and is on the endangered species list.

The black rhino has lived in Fish River Canyon before, and Natacha believes that it should be given a chance to return to the place where it once belonged. 'We try and introduce species that should be here or that have been here before,' she says, 'and just see how they go … and how they do.' If Natacha can reintroduce the black rhino to Fish River Canyon, she will have made a huge contribution to this species' conservation, and Natacha's commitment to these animals will have once again helped to save them.

Natacha is absolutely certain she's doing the right thing in repopulating Fish River Canyon and returning it to its natural state. She says that for her, it's partly about choosing her own way of life and partly about having a real purpose. She explains, 'It matters to me to know every morning that I wake up that I'm going to do something exciting, that I am where I am happy to be, that I live a way of life that I have chosen.' To this thought Natacha adds, 'It's wonderful. It's the dream becoming a reality in a way. It's the purpose of being here.' In working for the conservation of these beautiful and unusual animals, this young woman who grew up in a distant European city is realising her childhood dreams. At the same time, Natacha's animal rescue projects are helping her to find her true purpose in the vast wilderness of Africa.

After You Read

1. People killed animals in Fish River Canyon for which of the following reasons:
    A. eating, hunting, racing
    B catching, hunting, racing
    C. hunting, catching, sporting
    D. sporting, eating, hunting

2. Natacha's plan to 'bring nature back into balance' in Fish River Canyon means:
    A. She wants to stop hunting and killing.
    B. She wants to make a new place for animals.
    C. She wants to create a lush area again.
    D. She wants to build a zoo.

3. What is the writer's purpose on page 8?
    A. to introduce Natacha's leopard friend Chemun
    B. to present Natacha's love and commitment for animals
    C. to show Natacha taking action
    D. to indicate that animals love Natacha

4. In paragraph 2 on page 8, the word 'repopulate' means to:
    A. overlap the animals' territories
    B. restrain the animals that live there
    C. bring more animals to the area
    D. make the area popular once again

5. Why does Ulf have a feeling of satisfaction when he goes to Fish River Canyon?
    A. because he is very excited
    B. because he has done something to help
    C. because he is a vet
    D because he touched wild animals

6. In paragraph 2 on page 20, to what is 'their' referring?
    A. the Fish River Canyon animals
    B. Natacha and Ulf
    C. the giraffes
    D. the parks

7. Natacha has found a location that has _____ for the giraffes.
   A. everything
   B. something
   C. anything
   D. nothing

8. Why does Natacha seek Hermann's help?
   A. He is an ally.
   B. He is an excellent marksman.
   C. His speciality is catching giraffes.
   D. all of the above

9. Which is the most appropriate heading for page 36?
   A. Giraffes Escape
   B. Bouncing Around with Giraffes
   C. A Successful Relocation
   D. Workers Guide Giraffes to Trailer

10. The black rhinoceros will be an important addition to Fish River Canyon because it is an endangered species:
   A. True
   B. False
   C. Not in text.

11. On page 40, when Natacha says 'see how they go' means to:
   A. wait to see if they will succeed or not
   B. make sure they can run okay
   C. watch what they do every day
   D. discover where they hide

12. Which of the statements best expresses the writer's opinion on page 43?
   A. Natacha is a valuable part of the world.
   B. The modern world is going to affect all animals.
   C. One woman's dream can make a difference.
   D. One can find his or her true purpose in European cities.

Make Room for Animals

As the world's population grows, more and more wild animals are being forced out of their traditional habitats. Some endangered species are being killed for food or because they compete with humans for territory or food supplies, while others are valued for their fur or uses in medical research. This type of killing has completely wiped out some species of wild animals and is threatening many others. Fortunately, animal rescue organisations are working hard to preserve endangered animals such as elephants, jaguars and chimpanzees.

In Thailand, the Wild Animal Rescue Foundation of Thailand (WARF) is working to stop human-elephant conflicts in Prachuab Kiri Khan province. Much of the forest in this area has been cleared by farmers and is now used for planting crops, which has destroyed the elephants' natural habitat and caused a continuing food shortage for them. Elephant herds have begun to come out of the forest and feed on the farmers' crops, so the farmers often shoot or poison the animals. The Elephant Feeding Project aims to create an ecosystem that will allow animals and people to live together in peace. WARF gradually restores the forest areas, while teaching people who must give up farming other ways to earn a living.

In Central and South America, ecological groups are attempting to increase the number of jaguars, the largest cats found in the Western Hemisphere. Adult males can be up to 2.3 metres long and weigh up to 250 kilograms. There are still jaguars in the area near the Amazon River; however, the population elsewhere in South America is very small and the jaguar disappeared completely in the United States during the 1940s. At one time, the jaguar's skin was considered very valuable, but today laws preventing the trapping or shooting

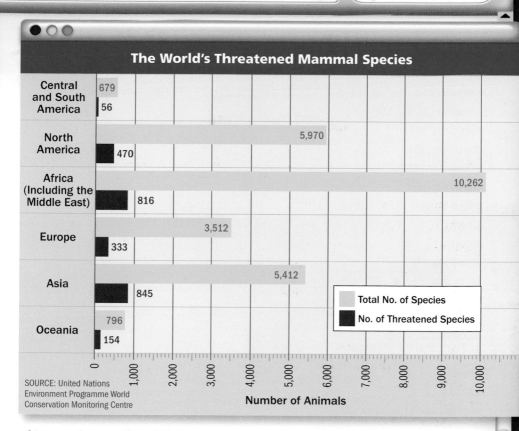

The World's Threatened Mammal Species

Region	Total No. of Species	No. of Threatened Species
Central and South America	679	56
North America	5,970	470
Africa (Including the Middle East)	10,262	816
Europe	3,512	333
Asia	5,412	845
Oceania	796	154

Number of Animals

SOURCE: United Nations Environment Programme World Conservation Monitoring Centre

of jaguars have reduced the demand for the skins. Many animals are still killed, however, because these predators often turn to cattle and other farm animals for food when their natural habitat disappears.

In the Bossou Hills of Guinea, West Africa, a Japanese scientist is working to save the area's rapidly shrinking chimpanzee population. The local population has been cut off from other groups because much of the forest connecting the Bossou Hills area with other forest regions has been cut down. Tetsuro Matsuzawa is working to replant groups of trees in the open fields to connect the forest areas once again. However, it may be several years before the world knows whether or not his plan will work.

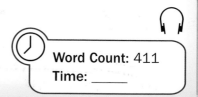

Word Count: 411
Time: _____

Vocabulary List

ally (11, 24)
black rhinoceros (3, 40)
bull (32)
canyon (2, 4, 6, 7, 8, 11, 12, 13, 15, 19, 20, 23, 25, 27, 35, 39, 40, 43)
cheetah (3, 11, 15)
dart (28, 33)
devotion (11)
ecosystem (12)
euphoric (36)
game animal (2, 4, 7, 8, 11, 19)
giraffe (3, 13, 16, 19, 20, 23, 24, 25, 26, 27, 28, 31, 32, 35, 36, 39)
leopard (3, 8)
livestock (4)
lush (35)
marksman (24, 28)
oryx (3, 4)
paradise (2, 7, 8)
poacher (4)
predator (7, 19)
pristine (2, 4, 12)
regain (one's) senses (32)
revert (12)
roam around (16, 19)
springbok (3, 15)
trailer (27, 32, 35, 36)
trophy (4)
vet (2, 11, 24)
wipe out (2, 7, 13, 19)